THE POETRY OF TENNESSINE

The Poetry of Tennessine

Walter the Educator

Silent King Books

SILENT KING BOOKS

SKB

Copyright © 2024 by Walter the Educator

All rights reserved. No part of this book may be reproduced in any manner whatsoever without written permission except in the case of brief quotations embodied in critical articles and reviews.

First Printing, 2024

Disclaimer
This book is a literary work; poems are not about specific persons, locations, situations, and/or circumstances unless mentioned in a historical context. This book is for entertainment and informational purposes only. The author and publisher offer this information without warranties expressed or implied. No matter the grounds, neither the author nor the publisher will be accountable for any losses, injuries, or other damages caused by the reader's use of this book. The use of this book acknowledges an understanding and acceptance of this disclaimer.

"Earning a degree in chemistry changed my life!"
- Walter the Educator

dedicated to all the chemistry lovers, like myself, across the world

TENNESSINE

Of Tennessine, where mystery beholds.

TENNESSINE

Born from the fusion of atoms' embrace,

TENNESSINE

In the cosmic symphony, where elements race.

TENNESSINE

Named after Tennessee, a land of charm,

TENNESSINE

In its atomic structure, secrets disarm.

TENNESSINE

Protons and neutrons, bound in delight,

TENNESSINE

In the quantum dance, where particles alight.

TENNESSINE

A fleeting existence, it swiftly fades,

TENNESSINE

In the cosmic ballet, where energy cascades.

TENNESSINE

But in its essence, a story ignites,

TENNESSINE

Of atoms entwined in cosmic flights.

TENNESSINE

A fusion of berkelium and einsteinium's touch,

TENNESSINE

Tennessine emerges, oh so much.

TENNESSINE

With an atomic number, one hundred seventeen,

TENNESSINE

In the periodic table, where elements convene.

TENNESSINE

In the heart of a collider, where particles collide,

TENNESSINE

Tennessine emerges, a cosmic guide.

TENNESSINE

Unstable and bold, it yearns to explore,

TENNESSINE

The depths of existence, forevermore.

TENNESSINE

In laboratories of innovation's might,

TENNESSINE

Scientists unravel its atomic flight.

TENNESSINE

Unveiling the secrets of quantum domains,

TENNESSINE

In the pursuit of knowledge's eternal reign.

TENNESSINE

Oh Tennessine, elusive and rare,

TENNESSINE

In the cosmic dance, you dare to compare.

TENNESSINE

To the wonders of creation, both near and far,

TENNESSINE

In the cosmic tapestry, where wonders spar.

TENNESSINE

With each proton, each electron's spin,

TENNESSINE

Tennessine weaves the fabric thin.

TENNESSINE

Of space and time, of matter and light,

TENNESSINE

In the cosmic theater, where all unite.

TENNESSINE

So let us cherish Tennessine's glow,

TENNESSINE

A reminder of the wonders below.

TENNESSINE

In the vast expanse of the cosmic sea,

TENNESSINE

Tennessine shines, for all to see.

TENNESSINE

In the dance of particles, where chaos reigns,

TENNESSINE

Tennessine endures, breaking cosmic chains.

TENNESSINE

With its fleeting presence, it whispers the tale,

TENNESSINE

Of the cosmic symphony, where wonders prevail.

TENNESSINE

ABOUT THE CREATOR

Walter the Educator is one of the pseudonyms for Walter Anderson. Formally educated in Chemistry, Business, and Education, he is an educator, an author, a diverse entrepreneur, and he is the son of a disabled war veteran. "Walter the Educator" shares his time between educating and creating. He holds interests and owns several creative projects that entertain, enlighten, enhance, and educate, hoping to inspire and motivate you.

> Follow, find new works, and stay up to date
> with Walter the Educator™
> at WaltertheEducator.com

www.ingramcontent.com/pod-product-compliance
Lightning Source LLC
LaVergne TN
LVHW051922060526
838201LV00060B/4131